Please visit our website, www.garethstevens.com. For a free color catalog of all our high-quality books, call toll free 1-800-542-2595 or fax 1-877-542-2596.

Cataloging-in-Publication Data
Names: Davies, Monika.
Title: Plants can poke! / Monika Davies.
Description: New York : Gareth Stevens Publishing, 2023. | Series: Plants with superpowers | Includes glossary and index.
Identifiers: ISBN 9781538279038 (pbk.) | ISBN 9781538279052 (library bound) | ISBN 9781538279045 (6pack) | ISBN 9781538279069 (ebook)
Subjects: LCSH: Thorns–Juvenile literature. | Spines (Botany)–Juvenile literature. | Prickles–Juvenile literature. | Plant defenses--Juvenile literature.
Classification: LCC QK921.D38 2023 | DDC 581.4'7–dc23

Published in 2023 by
Gareth Stevens Publishing
29 East 21st Street
New York, NY 10010

Copyright © 2023 Gareth Stevens Publishing

Editor: Monika Davies
Designer: Leslie Taylor

Portions of this work were originally authored by Devi Puri and published as *Plants that Hide*. All new material in this edition authored by Monika Davies.

Photo credits: Cover, p. 1 ahorizon/Shutterstock.com; Series Art (various accents) Ozz Design, Piotr Urakau, Designer things, Maxger/Shutterstock.com; p. 5 (top left) Cineberg/Shutterstock.com; p. 5 (holly) Stella Oriente/Shutterstock.com; p. 7 jurgal photographer/Shutterstock.com; p. 9 (top left) blueeyes/Shutterstock.com; p. 9 (desert) David ODell/Shutterstock.com; p. 11 (thorns) ImAAm/Shutterstock.com; p. 11 (spines) Prabhakarans/Shutterstock.com; p. 11 (prickles) Pavla Kafka/Shutterstock.com; p. 13 (roses) Mr.Boy/Shutterstock.com; p. 13 (thorns) Sandeep-Bisht/Shutterstock.com; p. 15 (suguaro cactus) Bernadette Heath/Shutterstock.com; p. 15 (barrel cactus) Randy Bjorklund/Shutterstock.com; p. 17, 21 (nettle hairs) Bildagentur Zoonar GmbH/Shutterstock.com; p. 19 (map) https://www.fs.fed.us/wildflowers/plant-of-the-week/Cirsium-discolor.shtml; p. 19 (field thistle) Thorsten Schier/Shutterstock.com; p. 21 (tea) Magdalena Kucova/Shutterstock.com; p. 21 (stinging nettle) nada54/Shutterstock.com.

All rights reserved. No part of this book may be reproduced in any form without permission in writing from the publisher, except by a reviewer.

Printed in the United States of America

Some of the images in this book illustrate individuals who are models. The depictions do not imply actual situations or events.

CPSIA compliance information: Batch #CSGS23: For further information, contact Gareth Stevens, New York, New York, at 1-800-542-2595.

CONTENTS

Prickly Plants 4
Thorny Defenses 6
Adapting to Survive 8
How Plants Poke 10
Pretty and Prickly 12
Cactus Jabs 14
A Thorny Tree 16
Don't Touch the Thistle! 18
A Stinging Poke 20
Glossary 22
For More Information 23
Index 24

Words in the glossary appear in **bold** type the first time they are used in the text.

Prickly Plants

Some plants have big, soft **petals** and leaves that are safe to touch. These plants seem to invite people to draw in close and smell their sweet blooms. But other plants have a thorny or prickly side. Come near them, and they will poke you!

Plants that poke come in many sharp shapes and sizes. They use different **structures**, like thorns, to **protect** themselves from harm. They will poke any enemy that gets too close—and that includes you! Let's learn more about these prickly plants.

Thorny Defenses

Plants use thorns, prickles, and spines to protect themselves. These are common plant defenses, or ways of guarding against an enemy. Thorns and spines are thin and sharp, and they look like something that can hurt you. They warn predators to stay away. Predators who **ignore** this message find out the hard way that thorns hurt.

Other plant defenses include poison and bad smells. Some plants **shrink** when they're touched. Others can change their colors, patterns, or shapes to blend in with their surroundings.

Adapting to Survive

All living beings, including plants, must find ways to deal with **threats** to their survival. Plants face many dangers, including predators, bad weather, and poor conditions in their environment, or natural surroundings.

Over time, plants developed adaptations, or changes that make them better able to live in their surroundings. Plant defenses, like prickles, are all adaptations. Long ago, plants that had these defenses survived longer than plants without them. These plants passed their adaptations on to the next **generation**.

How Plants Poke

Plants can poke using sharp structures. Thorns are sharp parts that stick out from a plant's branches or stem. Thorns are set deeply in the wood of the stem or branches. They are hard to cut or break off from the plant. Each has a stiff point.

Spines are thin, sharp structures. They are **modified** leaves, even if they don't look like most leaves. Spines are shaped to keep predators away and often help save water for the plant.

Super Fact! Prickles are small, sharp growths off a plant's "skin," or outside layer. Unlike thorns, prickles are not set deeply within the plant, so they are easier to pull off.

Pretty and Prickly

Roses are known for their sweet smell, beautiful blooms, and "thorny" stems. However, roses actually have prickles on their stems.

The prickles on a rose plant are small structures with sharp points. This defends the plants against herbivores, or animals and bugs that eat only plants. Animals might be drawn to a rose's sweet smell. Some think prickles on roses help keep predators away. This defense works on people too. Have you ever touched a rose plant? You have to be careful or you will get poked!

Super Fact!

Roses can have different kinds of prickles. Prickles on a rose's stem come in a wide range of shapes, sizes, and even colors!

Cactus Jabs

Cactus plants have a prickly **reputation**. These desert plants are covered with sharp spines, or needles. Cactus plants store water inside their stem, which thirsty predators would love to find. However, a cactus's sharp spines send a clear message: "Stay away!"

Cactus spines also serve another purpose. Plants lose water through their leaves. This is a problem for cacti living in hot, dry places. The plants adapted by replacing bigger leaves with spines. This keeps them from losing too much water in the hot sun.

Super Fact!

Cacti do not grow deep root systems. Their roots grow near the top of soil so the plant can get water from a large area when it does rain.

A Thorny Tree

Floss-silk trees are a sight to see. This tree has bright, star-shaped blooms and is covered in striking—and surprisingly sharp—prickles. These fat prickles have an edge that pokes anyone who comes too close! This tree first grew in a few countries in South America, but is now also found in some U.S. states, including Florida and Southern California.

The trunks of floss-silk trees are made of thin bark. Some think the tree's prickles help protect the trunk from animals that might climb up, like monkeys.

Don't Touch the Thistle!

Is it a garden plant or weed? The thistle plant is a weed to some, but others grow them in their home gardens. No matter where they are found, thistles are one prickly plant!

Prickles can grow on all parts of the thistle plant, including the stem and leaves. Prickles can even grow on the plant's flowers. The plant's edges send a sharp warning to animals to stay far away. Be careful if you come across a thistle. You don't want to be poked by its prickles!

Super Fact!

Some thistles are invasive species, or kinds of living things that spread quickly and are harmful when placed in a new area. This includes the Canada thistle and bull thistle.

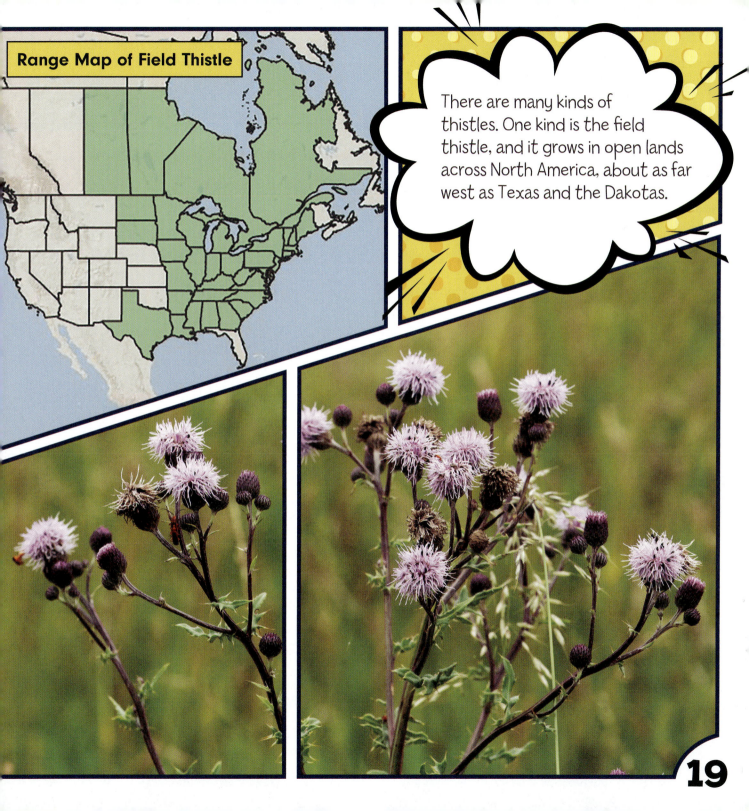

A Stinging Poke

The stinging nettle does more than poke its enemies. It stings! Thin, prickly hairs cover the plant's leaves and stems. If you brush against these hairs, they **inject** a stinging **acid** into your skin.

Stay on the lookout for plants that can poke or sting you, like the stinging nettle. Plants that poke want to stay safe, so they come armed with a sharp side. Their ability to poke is a helpful—but harmful—defense. It's best to stay far away from them!

super fact!

The acid from a stinging nettle causes our skin to itch, burn, and swell up. A rash, or a group of red spots on the skin, may also form.

GLOSSARY

acid: A liquid that breaks down matter.

generation: A group of living things born about the same time.

ignore: To choose to do nothing about something.

inject: To use something sharp to force harmful matter into an animal's body.

modified: Changed in some parts while not changed in other parts.

petal: One of the soft parts of a flower that is a certain color.

protect: To keep safe.

reputation: The views that are held about something or someone.

shrink: To make smaller in size.

similar: Nearly the same as something else.

structure: The way something is arranged.

threat: Something likely to cause harm.

FOR MORE INFORMATION

BOOKS

Loukopoulos, Beatrice. *Plants with Thorns, Spines, and Prickles.* New York, NY: PowerKids Press, 2019.

Schuh, Mari. *Prickly Plants.* Minneapolis, MN: Pogo Books, 2019.

WEBSITES

Cactus
kids.britannica.com/kids/article/cactus/352894
Learn more about cacti at this website.

Plant Defense
www.dkfindout.com/us/animals-and-nature/plants/plant-defense
Find out more about different plant defenses here.

Publisher's note to educators and parents: Our editors have carefully reviewed these websites to ensure that they are suitable for students. Many websites change frequently, however, and we cannot guarantee that a site's future contents will continue to meet our high standards of quality and educational value. Be advised that students should be closely supervised whenever they access the internet.

INDEX

adaptations, 8, 9, 14

agave, 9

cactus, 14, 15

California, 16

desert, 9

Florida, 16

floss-silk tree, 16, 17

flowers, 4, 12, 16, 18

holly, 5

honey locust, 7

leaves, 4, 5, 9, 10, 14, 18, 20

needles, 14

North America, 19

predators, 6, 9, 10, 12, 14

prickles, 6, 8, 10, 11, 12, 16, 18

poison, 6

rash, 20

roots, 14

rose, 12, 13

South America, 16

spines, 6, 9, 10, 11, 14, 15

stem, 10, 12, 14, 18, 20

stinging nettle, 20, 21

Texas, 19

thistle, 18, 19

thorns, 4, 6, 7, 10, 11